Black and White

An Account of the Root Cause and His Impact on the Universe

Ta Ha Publishing Unlimited

Contents

Foreword

Earthquakes, hurricanes, solar flares, hail, snow storms, and even fog, were once thought to be products of a planet perpetually in change, spewing out monstrosities that wreak havoc on earth, and in the souls of men. When it once was sufficient to call these events acts of God, the time has arrived with such virility and urgency, and on a scale that seemingly grows exponentially, that the above can be put into a unique category, never fully explained until now. There was a man, who promised a book, but left our secular plane of existence before it could be published; it is in his honor that I dedicate these words. And consequently, for reasons which I believe you will later understand, the words *entity*, *force*, *essence*, *the light*, and *creator* are all capitalized, and refer to the same One.

More than an Enigma

God is a combination of two things: that which He thought about, and that which He created. This means that the two are intertwined. Also, it means that His creation is responsive.

The question is raised often, why is it imperative that what He created heed His word? The explanation lies in D.N.A. sequencing and nutrient synthesis, where bodily chemicals are combined in a way rendering Him incredibility resilient and powerful, for He is from the very atom of the Universe itself.

Material manipulation, which collects the needed elements and minerals, is merely an issue of thought. For example, He causes His creation to draw upon each other, as He has ordained oneness. All things are responsive-from the ants to the trees-they all heed His call. Believe it or not, it was thought about from the very foundation of the Universe itself. This is why the Catholic doctrine of the Trinity places Jesus Christ in the Godhead.

It's not that he was actually there in the beginning of time, but that his body processed the same molecular and subatomic particles that were present in the start of all things. But once again, why would God make it imperative that His creation respond?

First, I must digress to a place more secure. His brain is similar to a magnet. In other words, it attracts particular thoughts that make up His overall being; the divine One. Immensely powerful concepts surround Him, and as a result, He controls mental patterns in human beings, as well as those on other planets. We simply must take into account the fact that astrophysicists have gathered substantial data concerning the heavenly spheres.

There is indeed a consensus that there is life in outer space. This raises other issues, such as His work and His relationship to mankind, which will be addressed later. This One chooses and cultivates men of a particular brain chemistry whose neuronal structure have an affinity, so to speak, with His.

The God Head: persons of a particular brain structure who surround the throne of greatness, who have inherited the kingdom once ruled by others of high intelligence.

There are many factors concerning His creation, and so, the divine paradigm continues on forever, enfolding in upon itself much like an Indian wigwam. To control one's thoughts is to harness God power itself. Deep conceptualization and inward belief dictates the outward figure one sees while viewing a mirror or a tranquil sea.

And said in another breath, the saga of man is written upon his chest. We all know of physical laws, and those same intricacies of a working system can be seen in Him, seeding and growing within the very fabric of the divine matrix. Further, His words and actions are calculated, resulting in right action, corresponding to the very positioning of His brain cells, rightly ordered to think correct.

And exactly so, as His stamina comes from within, born of neuronal transmissions, whereby He is pushed forward by an invisible force, bringing Him face to face with the manifestation of thought. Sustained stamina: the mirror of one's self, surrounding

and protecting Him from those events things that may be

encountered in His daily life. If people would only realize their

potential, life would be easier, for with suffering and pain is

relaxation and comfort.

Thought Manifestation

We continue on with a new realization which holds the key to His future events. Those things that He holds in an expectant way are bound to happen. He merely needs to decide upon that which He intends. Thought manifestations: His brain cells are groomed and groomed into the matrix, the God pattern.

He controls the outcome of future events, being guided into a

perfect path. Time tables and thought materialization are indeed

related with neuronal structure, and ultimately, related to the true

source of power and inspiration.

This leads us to another question. What is the origin of the

very first Essence? It must be defined as holy, because it is

undefiled and pure. I now direct the reader to the Biblical story of

Mount Sinai, where Moses stood near this Entity which was described as the holy flame, or as Jewish scholars call it, The Light. That which lies inside the seat of power is goodness, and only needs cultivation to render itself useful.

Similarly, the Force that motivates man is like a candle, which emanates strength and endurance, the eternal flame of the brain. There exist truth, never out of reach or far fetched, but stemming from within, eternally pulling on one's soul. Truly, to maintain the highest moral standards is the essence of integrity, and further, that which dwells within must be. The words of Moses in The Bible, "I am that I am," denote divine stamina, and in words sum up this train of thought.

A wondrous Creator; yet, should He decide to destroy the Universe, there is nothing to hamper such a decision. He turns the task over to man, and so is He chosen. The Supreme Being is created of infinite intelligence. One can view Him as the embodiment of the very first Essence. In the beginning, this Force made a motion towards that which it wanted: a determined movement of pure thought out of complete darkness.

And this motion was the beginning of the first record, or the book of life. Everything that was to happen is in His very creation, and subsequently, He has no beginning or ending. The very action of motion out of a black void is seemingly impossible, yet He chose for Himself light, and with it comes life.

This is why the loftiest characteristics are attributed to Him. He is success. All that He does is absolute, and hence, the Force is everlasting. He did not create death for Himself, only those things that He created, and so He is called God, and those who surround the throne of glory manifest certain facet or attributes of this most supreme One. He ordained for Himself continuity, and as a result, made His knowledge accessible.

Whatever being that came to exercise the most wisdom was allowed to rule for a certain period. Being filled with the One, He sits upon the throne as God the Most High. As mentioned before, only those things that He created die.

The ruling One, being made of mortal essence, would pass away, or would be taken down upon the appearance of One more powerful. Yet, the Force never dies or ceases to exist. The Force is unique among other powers, as it is an unwavering, exalted, all prevailing One, and its very existence is due to an inevitable motion towards life and light. The Supreme One: His total makeup is calculable, but only by Himself, or those created similarly, to such and extent that what He sees within His mind is ordered absolutely.

He cannot be overcome because He is the creator of all things. His pace is quantifiable, for He ascends to the throne of power much like a growth. Realities of His mind are realized in time, and moreover, can be plotted much like a matrix.

The mighty, powerful, self-created Being; nevertheless, this is not to say that humans were the first ones that He created. Everything is ordered, and creation is composed of periods of trillions of years. While the Bible says that God said "Be," and it was, some things took eons to come into existence. A human being is the most favorable manifestation of the Force. Still, throughout the trillions of years of life in the Universe, other beings existed, and may still exist.

For example, an angel, so much spoken of in eastern religions, was the first being that He created. The purpose of its creation was for it to enjoy the heavens, as the Lord at that time was not yet in human form; and so, with love, they were formed to enjoy His design.

Linkage of Like Minds

L inkage of minds can reveal many secrets of the heart. Due to neuronal structure of the brain, men are able to transfer knowledge to each other. Light increases the range of communication, as blood in the brain fuels the cells.

Thus, world plots are often seen by night, when circulatory fluids rise. Right thinking creates a circumference of power emitted by some. And so, to see the unseen is a matter of directed thought.

We must realize the eminence of right choice, which comes as a rush of information. His decision is destined to be realized, as the directed signal of His brain creates convulsions in the mind of men. Accordingly, physical manifestations of this occurrence is astounding, to say the least. Heavenly things await His command, a trillion years in the making. Indeed, the very Universe awaits His sway. Such an Entity is connected to an ordered affair, attended by beings similar, which regulate His growth if needed. They have supreme wisdom and power. Why a Godhead? What purpose does it serve?

The Elders, as those surrounding the throne of glory are called, have extraordinary mental and scientific capacity, and while knowledge is power, there is no way to measure their incredible gifts. These beings communicate with each other by brain linkage, sending both image and sound to one another. This is but one example of their tremendous capabilities.

So by realizing that God assumes the form of man on Earth, we close the lid on religious bigotry and ignorance. How does God Supreme fit into all of this? As a species, we have come to believe that God is force and power. This is surely true, as those neurons of His brain vibrate with frequencies that collectively form thought.

He is the One on earth whose brain structure is most profound in its complexity. When the neuronal transmissions match those of the receptor individual, He is able to dictate action within that person through linkage. Whatever man attempts, he must eventually fulfill; he can only have what he strives for. The brain is such a great creation, with neurons aligning themselves in such a fashion that a matrix is established, and thoughts are easily manifest.

The Sun was created in a similar fashion, as He realized that light could power, and would subsequently result in a harmonious system. Could it be said then, that the Sun, Moon, and stars act upon a wish?

It is the natural order for such things to take place, and therein lays the gift for all mankind. Glorious is the child who captures, in essence, the notion that all manifestations of his immediate environment are merely time capsules in the making;

that the inner workings of His mind has produced such, and brought on only by tantamount principles of brain physiology. Therefore, it can be safe to deduce that the possessor of such neuronal structure as described is He whom we call God.

This conclusion is based on several facts. One is that His thoughts are beyond those of ordinary men, rendering the manifestation of His will astounding. He accomplishes great tasks with mere thought; easily and inevitably.

This neuronal arrangement spoken of gives Him those attributes and powers that people have attributed to the Lord. As previously stated, He connects with other brains according to their chemical composition. Telepathy is a gift possessed by such a One and in a way yet to be explained, other minds are directly manipulated by this most powerful Entity. It should be said, therefore, that body chemistry weighs heavily upon the brain's physical structure and operation.

It is understood that certain materials influence the brain's neurons in such a way which inevitably will dictate what kind of thought is formulated. The overall mental state is part of this complex picture as well. Naturally, we turn our attention now to emotions. Are these feelings entertained by One of such power? He most likely had a formative stage, and would have been subject to stress, anger and pain-all the emotions of a living soul.

The Kingdom

God, the all powerful One, has endless ways to influence the natural world. He sends winds, which represent several forms of light, violence, and thought control, which are fueled by the predicaments men make for themselves. Chicanery and greed, for instance, propel these unusual forces, which are alive

nevertheless, who die and are reborn just as any creature. The

generative power they hold blow away and subsequently destroy

other negative spiritual traits such as avarice and evil, as they

contain water and oxygen, components of life itself.

The color of the winds is opaque, characterized by the

blankness of the Universe from which they are created, and as the

heaven was void and without form, they take on their Lord's color, which changes at a moments notice. Surely, He has power over the natural world itself. Solar flares, for instance, are a mystery, and far from the intellectual grasp of those who study the complexity of this issue, in the sense that those occurrences on the Sun reflect the state of man on earth, each one being reciprocal of the other. The Sun, the most misunderstood thing in our solar system, is presently undergoing a change, as is man and the Earth, being fitted to exist in a new universal environment, the process of which is ongoing, and of which the final state is know only by God. The solar

eruptions emitting charged particles are signs for men to reflect,

that they may ponder the vastness of created things, the

consequences and repercussions of the complexity of our galaxy,

and so that leaned men can calculate the apparent parabola of

thought/matter.

Yet, it is not merely will that is causing these eruptions we

currently see bombarding earth, but it is the compounded reality due to the existence of the One. Hurricanes, for example, are unique creatures, which are instinctively called by He who is most intimately connected with the heavens and earth. His affinity with nature is so profound that these circular winds (called cyclones by some) gather and await His command.

This could be near the secret of the cosmos: that prized jewel sought after by meteorologists, who marvel at the apparent dichotomy which is the Universe. The destruction associated with its winds is part of a domain of power, and, being a living creature, it too needs food, which is the warm water it inhabits. The dark whole witnessed in its center is similar to a brain stem, as it were, with its neuronal system like the outer bands that whirl around, feeding upon warmth and moisture, subsequently giving it the appearance of a highly charged storm system.

On a smaller scale, tornadoes are also atmospheric creatures, akin and often spawned by its larger cousin, who search for favorable air currents, bobbing and weaving dangerously through areas gulping in that which fuels it.

However, to say that hurricanes and tornadoes are living is to suggest that they have a mind of their own, and this deserves a thorough explanation. But fathom this: the electricity and water contained in these giants constitute two of the most important components needed for life. And while these things do not fit the description of animal life as we know, they are real live creations nevertheless.

He even affects the physical Earth. Earthquakes are caused by the One described, and were people to objectively take into account these words; they would come to additional realizations. Possibilities and equations compound by the second as we take a comprehensive look at the importance of this mighty One.

Carefully, you are guided into realms uncharted. Our very earth feeds and consequently reacts upon the tremors of His soul, with its physical equivalent being the actual mechanism of His universal gravity which ultimately causes the geologic equivalent. The exact crustal movement, or earthquake, is not presently able to be plotted by geologists, who are coming to a higher understanding of the natural world, which is the fact that earthquakes have a relationship with the rotation and speed of the Earth.

The missing piece to this enigma is the presence of the One, whose very existence has a bearing on geologic processes, in ways yet to be explained. Universal gravity is behind many of the phenomena we see happening on earth and in outer space, and is something that physicist have deduced is present in our solar system, due to the accelerated motion of the heavenly bodies, seen most plainly in the unusual activity of the Sun. But here is the big question: How on earth can One have power to such an extent that even the jet stream conforms to the force emanating from Him, and

further, cause the Sun to emit potentially harmful gamma rays

earthward, disrupting the countless mechanical devices on our

planet?

Universal gravity is an example of the timeliness of this

One, in the sense that the force emanating from this Entity is that

explosive charge akin to a thought by God Himself. Finally, the

Moon, that huge piece of real estate above our head is undergoing a metamorphosis, due to a thought by this powerful One. It merely

awaits the holder of universal gravity to make it complete. Sorry

for losing you all, but this is a story trillions of years in the making,

finally unfolding before us.

The Moon must be charged, however, and this remains a

feat far beyond present scientific explanation. What is a charged

Moon? A stimulated Moon centers on a working atmosphere,

stemming from the surrounding accelerated atoms which interact with dark matter. This will in turn start a chain of events

culminating in the first rain drops to hit the surface of that huge

rock, which will be a miracle placed with other magnificent

accomplishments recorded by writers of the past.

For we know that truth is like a seed, which grows into a

tree, who's branches we depend on for our everyday lives. In the

follow up book, I will dwell deeper into the specifics of how He

affects our world. I am reminded of a scripture which goes

something like this: "On the day that the heaven is cleft asunder,

so it will be frail that day". Simply, His presence and global

consequence propels us to consider the spiritual and religious

implications of this most powerful One.